DR
English Books

Cover: Image by Greg Egan

Translation from Italian by J.C. van den Berg

2014 Second reprint
First published in Italian: 2004
Viale Manzoni 59
00185 Roma
Tel. +39.06.77 20 90 20
Fax +39.06.70 47 40 67
E-mail: direnzo@direnzo.it
Internet: http://www.direnzo.it

Carlo Rovelli

What is Time?
What is Space?

Di Renzo Editore

Introduction

I have devoted a large part of my life to scientific research. But this has been a rather late-born passion. As a young man, the entire world interested me, but not science in particular.

I was born and grew up in Verona, in a quiet family. My father, a man of rare intelligence, was an engineer who ran a small business. I got from him the pleasure of trying to understand the world in an intelligent way. My mother, a real Italian mother full of excessive love for her only child, helped me with the "investigations" I did for elementary school, and stimulated my curiosity for discovery and learning.

I attended the classical lyceum of Verona. We studied more Greek and Latin than mathematics. The school offered rich cultural incentives, but it was pretentious and provincial, closed in its attempt to safeguard the privileges and the identity of the small provincial bourgeoisie. Several of the teachers had been fervent fascists before the war, and had remained so in the secret of their heart. It was the sixties and seventies, decades of conflict between generations. The world changed rapidly and most adults had difficulties to accept the changes. They dug themselves in, around defensive and sterile positions. I had little trust in the world of adults and even less in my teachers. I clashed frequently with them and with all figures of authority.

My adolescence became a period of revolt. I did not recognise the values expressed by the world around myself. It was also a period of great confusion in which nothing seemed to be certain anymore. The only thing that seemed clear to me was the feeling that the world I saw was very different from what seemed just and right to me. I wanted to become a bearded vagabond, to stay away from a world that I did not like. But I also read voraciously and the books were telling me about other worlds and different ideas. I felt there were wonderful treasures hidden in every book I had not yet read.

During my University years in Bologna, my own confusion and my own conflict with the adult world made contact with the common course of the generation to which I belong. It was the time we all wanted to change the world. To transform it into a more gentle and a less unfair place. There was a desire to try new forms of living and loving. We experimented with new ways to live together, we wanted to try everything; we travelled on the road, with the mind, and with our dreams, we endlessly fell in love and endlessly talked. We wanted to learn to see things with different eyes. Sometimes there was great distress, other times we felt at the dawn of a wonderful new world.

We lived on dreams. We travelled a lot in search of new friends and new ideas. At twenty, I decided to go on a long trip by myself, to wander around the world. I said I wanted to look for adventure and to seek for truth. Nowadays, approaching fifty, I think that it was a good choice. Obviously, I smile about my ingenuousness at the time, but I feel I am still following an adventure that began in those days. The path has not always been easy, but the foolish hopes and the unbridled dreams have not deceived me: it was sufficient to have the courage to follow them.

With a group of friends we set up one of the "free radios" of those days, *Radio Alice*, in Bologna. It was an open microphone for everybody, where experiences and dreams were exchanged. With two of those friends we wrote a book telling the story of this Italian rebellion of the end of the seventies.

But the dreams of revolution died quickly and order got the upper hand; one does not change the world so easily.

Thus halfway through my university years I found myself even more confused, saddened by the sensation that dreams I had shared with my contemporaries of half the planet were already fading, and with no perspective on what to do next with my life. The idea to join the race up the social ladder, a career, to earn a bit more money and grab a crumb of power seemed too sad to me and I did not feel up to it. But there was still the world to explore and behind the hills I imagined always boundless horizons.

Scientific research has come towards me at this point as a boundless space of liberty, and an extraordinary adventure, old and new. I studied because I had to do exams, and to delay the compulsory Italian military service (at the end I just burned the draft card), but the things I studied aroused my curiosity more and more and fascinated me ever more. In the third year of the undergraduate Physics program, one meets the "new" physics of the twentieth century: quantum mechanics and Einstein's theory of relativity. These are extraordinary conceptual revolutions in which the view of the world profoundly changes. Earlier ideas are challenged, even the most revered ones. One discovers that the world is not at all as one thought it to be, one learns to see things with different eyes. It was an extraordinary mind travel.

I slid from a cultural revolution into a revolution of thought, almost without realising it.

I discovered a kind of thinking, scientific thinking, which establishes rules to understand the world, and then challenges them and changes them. This freedom, this free path to the pursuit of knowledge, fascinated me. It was just curiosity and perhaps what Federico Cesi, the friend of Galileo who had the vision of modern science, called "the natural desire to know", which made me find myself, almost without noticing it, immersed in the problems of theoretical physics.

The interest for physics was thus born more by accident and by curiosity than by a conscious choice. As a young student I was good in physics and mathematics but I was much more interested in philosophy. At the University I had enrolled in Physics and not in Philosophy mainly because in my naive mistrust of established institutions I considered philosophical problems too serious and important to be willing to discuss them at school.

When the dream of constructing a new world ran around against reality, I fell in love with science, where there are infinitely many new worlds still to be discovered, and where it seemed possible to follow a free and shining path of exploration of what surrounds us.

Science has therefore been a compromise for me: the possibility of not giving up the desire for change and adventure, to maintain my freedom of thinking and to be what I am, without too much into conflict with the world around me; on the contrary, by doing something which the world appreciates.

I believe that much intellectual or artistic work is born from this conflict: it is a kind of refuge for potential misfits. Still, society needs this kind of people. Our society lives in a state of equilibrium. On the one hand there are the forces that assure its stability and permanence and that prevent disorder from ruining what has already been acquired; on the other hand there is the inexhaustible desire for change and justice which tends to displace it, to make it go forward and to make it evolve. Without this desire for change, civilisation would not have grown and would not have reached the point where it is now; we would still be adoring the pharaohs. I think that the wish for change of the young, which always returns even if it is sometimes slumbering, is the first source of the evolution of society. If on the one hand society needs people of order who safeguard stability but rein in history, on the other hand it needs people who live of dreams and launch themselves into discovering new worlds, new ideas, new ways of looking at things and of understanding reality. It is the people who lived dreams who in the past have been able to think and construct our world, and only from many new dreams can the future be born.

The discovery of a problem: quantum gravity

In the fourth year at the university of Bologna, a paper fell into my hands, written by an English physicist, Chris Isham, in which the author spoke of "quantum gravity". The paper explained that at the basis of contemporary physics lies a fundamental unsolved problem. It has to do with the definition of space and time and the basic structure of the world. The problem is the following.

The great scientific revolution of the twentieth century consists of two parts. On the one hand there is *quantum mechanics*, on the other hand there is Einstein's *general relativity*.

Quantum mechanics, which describes well microscopic objects, has changed profoundly what we know about matter. General relativity, which is a theory explaining the force of gravity, has profoundly changed what we know about space and time. Both theories are extremely well verified, and both are at the basis of much of contemporary technology.

However, the two theories appear to be incompatible. They lead to two very different ways to describe the world. Quantum mechanics, in its present formulation, uses the old notions of space and time, which are contradicted by general relativity. General relativity, in turn, is formulated using the old notions of matter and energy, contradicted by quantum mechanics. We do not know how to merge the two into a unique consistent conceptual framework.

We are therefore today in a situation where we have a general description of the physical world which is badly fragmentary and internally inconsistent. Today, in reality, we no longer know very well what space is, what time is and what matter is. The problem of reconciling these two visions of the world is the problem of *quantum gravity*. It is a fundamental problem of greatest importance at the basis of contemporary physics, and there are many groups of scholars in the world that are seeking to solve this problem.

Obviously it is a difficult problem. But with the rash courage one has only when one is twenty, I immediately decided that this was the problem I wanted to work on. I was attracted by the idea of studying the concepts of time and space, and by the fact that the situation appeared to be so confused. In Italy almost nobody was working on this problem. My professors strongly and unanimously discouraged me to go in this direction. They told me: "You will never find a job" (which is what I now say to the young people who want to study these things); or, "You have to join a powerful and already well established group", etc. Fortunately for the young people, the only result of the cautious advice of adults is often that it only strengthens their cheerful stubbornness.

Space and fields

It is not easy to illustrate theoretical physics to someone who is not fully immersed in it. But I will try to explain how the problem of quantum gravity arises and what it has to do with the structure of space and time.

Let me start with space. The common idea we have about space goes back to Newton. We can think of space as of a *table* on which all the facts of the world occur, or a great *container*, inside which everything happens. For example, the particles, subjected to forces, move around on this table-space. The Newtonian picture of the world is that of this table-space, where particles run around, pushed and pulled by forces.

At the end of the nineteenth century Maxwell and Faraday change this view of the world a little, by bringing in a new ingredient. This new ingredient is the electromagnetic *field*. The electromagnetic field is the entity that produces the electric and magnetic forces. It is an entity which is spread out and fills all space. Faraday described the electromagnetic field as a collection of *lines* originating at positive electrical charges, ending into negative electrical charges, and occupying space everywhere. At each point of space there is one of these lines passing by. The direction of the line at the point is a vector (a small arrow), tangent to the line. If there is a charge at this point, the charge will feel a force in the direction of this vector. See Figure 1.

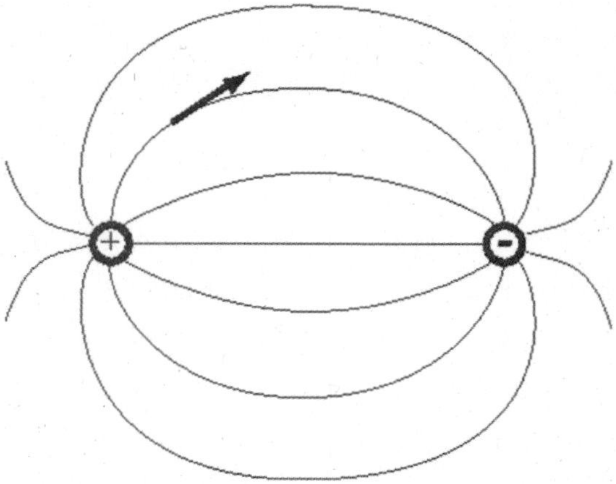

Figure 1: The electric field around two charges; the field consists of lines, the Faraday lines.The direction of the field at a point is indicated by the arrow.

The great discovery of Faraday and Maxwell is that these "Faraday lines" exist also in the absence of charges. If there are no charges where to begin or end, the Faraday lines form closed curves in space. These are called *loops*. One of those loops, or closed Faraday lines, is shown in Figure 2.

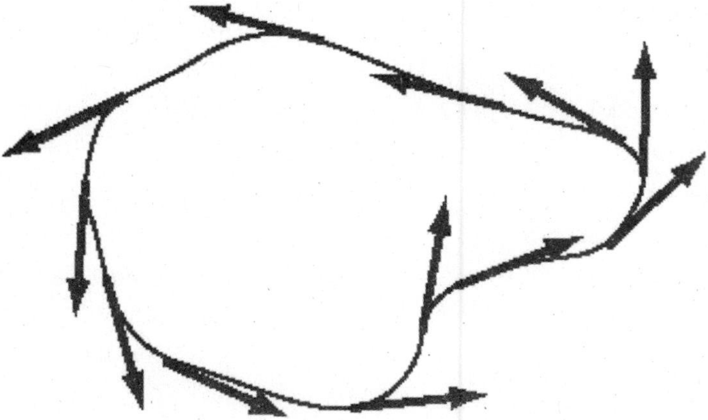

Figure 2: A closed Faraday line, i.e. a *loop*. The arrows represent the electric field, which everywhere is tangent to the Faraday line. These lines fill all of space and constitute the electromagnetic field.

The Faraday lines move, like waves of the sea, under the influence of the charges and of one another. Maxwell has formulated the equations that govern this movement of the Faraday lines. A stroke of genius of Maxwell is to have understood that light is nothing else but the fast undulatory movement of those lines. In fact, people often say that the fields are "invisible": nothing more false, because we see nothing but the fields! Indeed, we see the light reflected by the objects, not the objects themselves: what we see is the field and its undulations. The discovery of the fields, by Faraday and Maxwell, has opened the door to modern technology such as radio, TV and electronics, which has changed our lives. But the modification of the Newtonian picture of the world has not been major: there is still the space-table, with things moving on it. The novelty is that there are fields in addition to the particles.

General relativity

The true revolution in the picture of the world arrives in 1915, with Einstein. First, Einstein, understands that to describe the force of gravity (which pulls us down and which keeps the earth in the vicinity of the sun and the moon in the vicinity of the earth) it is necessary to introduce a gravitational field similar to the electromagnetic field. In addition to the Faraday lines of the electromagnetic field, therefore, there are also Faraday lines of the gravitational field.

But the great discovery of Einstein is something else: he discovers that this *gravitational field* and the *table-space* of Newton are one and the same. Space, which Newton had described as a fixed table, in reality does not exist. What does exist is the gravitational field: a physical object that can move and undulate like the waves of the sea. The force of gravity is the result of the effect of this undulating field, or of this undulating space, upon any object immersed in it. Space and gravitational field are the same entity: from then on, when one says space one also means gravitational field, and when one says gravitational field one also means space. It is an unexpected and spectacular discovery.

Hence, the world is not made of particles and fields that live in space, but of particles and fields, and this is all. Particles and fields do not live in space: they live on another field, i.e.

fields live, so to say, on top of one another. We live on the gravitational field or within the gravitational field, not on a fixed table-space.

Imagine looking at an island in the ocean. On top of it there are many animals: we say that we have seen "animals on an island". Then a young marine biologist called Einstanium investigates this a little more carefully and discovers that the island isn't an island: in reality it is a big whale. Thus, the animals don't live on an island but on another animal. No longer "animals on an island" but "animals on animals".

The discovery that the island is in fact a whale reveals to us that it is an animal like the others and therefore there aren't two entities of a different kind: animal and island; but rather only entities of the same kind, animals, which live "on top of one another", without the need to have dry land to put their feet on. In the same way, Einstein understood that the fields don't need to live on a fixed table-space, because they can live "on top of one another". After Einstein's discovery one no longer thinks of "fields in space" but of "fields on fields".

The world of Newton is like the island with animals on top of it, that is, consisting of the table-space which is fixed, static, an immobile base, and very different from the fields and the particles which move around on top of the table-space, like the animals move around on the island. Einstein discovered that Newton had made a mistake: table-space is not at all something that much different from the fields and the particles that move on top of it. On the contrary it is itself a field, just like the others. It may move, ripple and fold, its behaviour being subject to equations (which today are called the Einstein equations), just like the electromagnetic field.

This is Einstein's theory which is called general relativity. It is called "relativity" because in the theory there no longer exists an absolute location of things in space, but only the location of physical objects with respect to one another, that is "relative" to the one another. It is called "general" because though the theory is born only as a theory of the force of gravity, and thus to explain why and how things fall, its importance is "general" in the sense that the theory changes the notion of space and thus completely changes our entire understanding of the physical world.

This theory is very beautiful. It is not easy to study because intricate mathematics is necessary to formulate it with precision: the mathematics must describe fields which live upon themselves and not on the table-space. But when one understands it well, one can only be fascinated by its simplicity and its conceptual clarity. Concepts initially completely disconnected - space, the force of gravity, the fields - all become aspects of one simple entity: the gravitational field.

Until ten or twenty years ago general relativity was considered a very exotic theory having few applications and few experimental confirmations. In recent years, however, there has been an explosion of experimental confirmations and applications of general relativity. Applications are found in the most diverse fields: from astrophysics to cosmology, and to the experiments exhibiting gravitational waves, that are the vibrations of the Faraday lines of the gravitational field, predicted by the theory.

From the many predictions of the theory that are now confirmed in a spectacular way I mention only the existence of black holes, whose existence is nowadays well established: a number of black holes have been identified in the universe. From the applications, let me mention the GPS (the Global Positioning System): this little device, which is mounted on some cars, or that one buys in sports shops, gives the exact location on the planet, and could not function without taking general relativity into account.

Quantum mechanics

The other great revolution in the physics of the twentieth century has been quantum mechanics, which has changed our way of thinking about matter. Quantum mechanics, which I touch upon only very briefly, fundamentally tells us two things. First, many physical quantities have a granular or, as one says, "quantized" structure. The electromagnetic field for example, if observed at a small scale, is made of grains or "quanta" that are called photons. In the same way some quantities, for instance the energy of an atom, can only have certain particular values (the "energy levels" of the atom), which can be calculated from the theory. It is as if the energy of an atom is also granular.

Second, the dynamics is not deterministic but probabilistic. This means that the way in which things move is determined by probabilistic laws: it is impossible to predict the future with certainty, but we can calculate very precisely the *probability* that something will happen, i.e. the number of times it will occur if we repeat the same physical situation many times. For example, a particle is not described by its position but rather by a "cloud" of probability which represents the probability of any position in which the particle may be found.

Thus, quantum mechanics is the discovery that at small scale the world is granular and non deterministic.

In spite of this strangeness, quantum mechanics is a very mundane theory, which is at the basis of a large part of modern technology. Computers' hardware, just to make one example, is built using quantum mechanics routinely.

Quantum gravity

I can finally arrive at the problem I meant to illustrate, namely the problem of combining what we have learned with general relativity with what we have learned with quantum mechanics.

If one now combines the basic ideas of general relativity and quantum mechanics, it follows immediately that since space is a field (the gravitational field) space must have a granular structure, as does the electromagnetic field. The quanta of the electromagnetic field are the photons. The quanta of the gravitational field must be "grains of space", because the gravitational field is the physical space. The dynamics of these grains must be probabilistic. Hence space (i.e. the gravitational field) must be described as "clouds of probability of grains of space".

But what does it mean to say "grains of space"? How can we describe them? By what mathematics? What are the equations? What does it mean to say "clouds of probability of grains of space"? What consequences does it have for what we can observe and measure? How can we imagine and interpret a world described as clouds of probability of grains of space?

The problem of quantum gravity is how to construct a mathematical theory describing these probability clouds of grains of space, and to understand what it means.

However, the problem does not end here. As I will explain in more detail later, another great discovery of Einstein, made in 1905, that is ten years before general relativity, was that space and time can only be described *together*. They form an unbreakable whole called spacetime. Hence, it is spacetime which becomes granular and probabilistic, not just space. Therefore it is necessary to construct a new conceptual scheme which allows us to think about the world when *time* as well is no longer a continuous variable flowing on its own, but becomes part of this probability cloud of "grains of spacetime".

This is the extraordinary open problem, the existence of which I discovered in my fourth year at university.

Thus, while on the one hand I was writing my book on the students rebellion (a book that the police did not like and that earned me a good beating up in a room at the police office of Verona: "Tell us the name of your communist friends!"), on the other hand I immersed myself more and more into the study of space and time and into the mystery of quantum gravity.

I succeeded with no small difficulty in entering a PhD program in Padova. I chose as my supervisor a professor who allowed me to continue to study what I wanted. I devoted the years of my thesis to study in a systematic way everything that was known about the problem of quantum gravity and all the attempts existing at that time to address it. The other graduate students already published their first papers. I spent my graduate years without producing a single publication. What interested me was to study, to understand, not the possibility of a future career.

Twenty years ago very few solutions, even only embryonic ones, were known. The route that seemed most promising had been opened by a great figure, John Wheeler.

As a young man John Wheeler had been a collaborator of Niels Bohr one of the early giants of physics in the twentieth century. He had participated with him in the birth of quantum mechanics and later had occupied himself with nuclear physics and was among the creators of one of the first models of the atomic nucleus. During the war he was in the United States where he played a central role in the grim and tragic events related to the atomic bomb. It was during a discussion in his very office that it was decided to write to Roosevelt that one could and should make the atomic bomb because one feared that the Germans could make it first, which later turned out to be unfounded. After that Wheeler had been working on gravity and had become the principal collaborator of Einstein. It was he who introduced the felicitous term "black hole" and amongst his students was Richard Feynman, perhaps the greatest physicist of the second half of the twentieth century. In short, he has been a protagonist of the highest rank in the whole development of twentieth century physics.

Wheeler has had a series of very important intuitions and ideas that are at the basis of research into quantum gravity. He suggested that spacetime at the smallest scale is a kind of fluctuating foam (a spacetime foam). Together with Bryce DeWitt, another great American scientist, Wheeler formulated an equation that should be the basis of quantum gravity. However the equation, called the Wheeler-DeWitt equation, presented many difficulties: the mathematics on which it was based was not well-defined and its physical significance was very obscure.

This was the confused situation I encountered in my graduate years.

Nowadays, twenty years later, the situation is different. Some possible solutions of the problem of quantum gravity are known, although one does not yet know which is the right one. It has been my fortune to find myself in the situation of being able to participate in the construction of one of these possible solutions: *loop quantum gravity*.

Loop theory

My next step was to start travelling again as when I was a boy, still to seek new ideas and new friends, but this time with a precise aim: to meet people who were interested in quantum gravity, in the problems of space and time. I set off to visit and talk to the greatest figures in the world of quantum gravity, using money obtained from several sources: some money was assigned by Italian law for graduate students to go abroad; some came from a private scholarship, made available by the Fondazione della Riccia (the existence of which I had discovered only due to a notice on a board of the Department of Physics in Trento); some money I had been able to save myself. I announced my visit by letter (there was no e-mail yet) and I went to visit them.

London and Syracuse

Chris Isham, the author of the article that had stirred my enthusiasm for the problem, was the first I went to see. I stayed two months with him at the Imperial College in London. At Imperial, I met for the first time the colourful and international world of the theoretical physicists: young scientists with a tie and a suit mixed in the most natural manner with others with bare feet, long hear and bands on their forehead. All sort of languages and faces, all sharing the same respect for curiosity and intelligence. There was a sort of secret joy of the difference. There was the same free and joyful spirit of the hippies communities I had met and loved in my travelling of a precedent era.

Chris was the guru of quantum gravity. He knew everything one could know about it. He also knew Jungian psychoanalysis, theology, and all sort of other subjects, which he interwove naturally in his conversation. He had a soft and gentle charm, half old wise man, capable of giving the right advise to each of us, half eternal boy, always in awe for the wonder of the world. I explained him my first confused ideas and listened to him. I brooded over what he told me during long walks in Kensington garden, the magical garden of Peter Pan, the boy who did not want to grow, which is just near Imperial College. I photocopied everything which there was on the subject at the College and read enormously.

One day, Chris told me about a young Indian scientist in America, Abhay Ashtekar, who had been able to rewrite Einstein's general relativity theory in a slightly different form, simplifying the problem. According to Chris, to approach quantum gravity starting from the new formulation of Ashtekar might make research more easy.

I therefore went to Syracuse University, where Abhay Ashtekar worked. It was Syracuse New York, in the USA, not Syracuse in Sicily. Still, to go to a city with the same name as the one of the greatest scientists of all times, Archimedes of Syracuse, sounded as an auspicious sign to me.

I stayed there two months, to study Ashtekar's new formulation of general relativity, which had not yet been published. Abhay was shining with energy. He already had a small group of young scientists around him, which he directed with the fascination of his personality, meticulous and courageous at the same time. We all met in a room and he filled blackboards with his tiny and accurate writing, making again and again the "point of the situation" and the "list of open issues". His way of thinking was analytic: he kept coming back on the same argument over and over, to revise it, to correct it, to reformulate it, until a small crack, so far invisible, begun to appear, and a novel direction, so far hidden, was unveiled. He did not accept mistakes, or zones of shade, in his thinking. He seemed to represent a sort of magic equilibrium between East and West, one of those novel form of intelligence that are borne when different civilizations have the courage to merge. I participated to these meetings eager to learn.

In the meantime I had written my first, not very interesting papers in physics, and I went, uninvited and on my own account, to the conferences in which the subject was discussed. At one of those conferences, in Santa Barbara, California, I came to know of the existence of a young American scientist, Lee Smolin. He used the new formulation of Ashtekar and working together with a friend, Ted Jacobson, had succeeded in finding some strange solutions of the Wheeler-DeWitt equation. I therefore went to see Lee at Yale University to learn what these solutions were like. It was the beginning of a great friendship.

Yale

The day before I left for Yale, my fiancée of those days left me. This threw me in the darkest desperation. I was not in the right mood to go to Yale and talk physics and I almost decided not to leave; but I could not postpone it any longer, and went anyway. When I arrived in Lee's office, I started to talk somewhat timidly about my studies. But I couldn't avoid thinking of my failed love, and tears came to my eyes. Lee was startled, but when I explained, apologising, my unorthodox behaviour, he mentioned his own fiancée, recently gone, and he invited me to go sailing with him. We forgot physics and spent the afternoon sailing and talking about love, life, and dreams.

During the next days, Lee begun telling me about the confusion in his attempts to understand the new solutions of the Wheeler-DeWitt equation found with Jacobson. Lee's way of thinking was the opposite of Abhay's: Lee only looked ahead, never back. He tried to see across the obscurity of what we do not know. Trying to guess what could exists behind the screen of our immense ignorance. He had absolutely no fear of saying silly things: he seemed to think that one right intuition is well worth a hundred mistakes. Lee is a visionary, a kind of Giordano Bruno, the renaissance thinker who first envisioned an infinite space filled with infinite worlds, or a kind of Kepler, the first who freed the planets from their crystal spheres and left them free to follow mathematical lines in space; men who have been able to dream new ways of conceiving reality.

The strangeness of the solutions found by Lee and Ted consisted in the fact that every solution depended on a closed curve described in space, a ring, a *loop*. What did these loops mean? In the course of a long nightly walk on the campus of Yale, while discussing the problem over and over again, we saw a possible solution. These loops are the single Faraday lines of the gravitational field. They are *individual* lines, instead of a *continuous* of Faraday lines, as for the classical fields, because this is quantum mechanics, where continuous is replaced by a discrete entities. But since the gravitational field is space, we must not think that they are loops immersed in space: they themselves are space! It is space itself which is constituted by these loops.

This was what the equations were telling us. From those conversations, a new idea was born, which would lead to the theory today called *loop quantum gravity*, and which is considered one of the ways to solve the problem of quantum gravity.

For several weeks we worked frantically, rewriting the whole Wheeler-DeWitt theory entirely in terms of our loops. We succeeded in obtaining an equation which was better defined than the original Wheeler-DeWitt equation, we found a great many solutions and began to understand their meaning.

The solution determined by a single loop represents a universe consisting only by a thin ringlet of space and nothing else. In other words, space without these "loops of gravity" does not exist, because the loops themselves and their relations constitute space. The existence of these universes made of a single loop was the first concrete indication of the quantum granularity of space.

It was a marvellous time. In the following weeks we flew to Syracuse, to Abhay Ashtekar, then to London to Chris Isham, and then to a big physics conference in Goa, India, to announce our results. We drew attention and positive reactions from a good part of the scientific community.

Rome

The following years were devoted to the development of the theory. I had finished my Ph.D. thesis. I had obtained a scholarship from the Italian INFN (Istituto Nazionale di Fisica Nucleare). Since I was not connected to any research group, I could use this scholarship where I wanted. I decided to go to the University of Rome, La Sapienza ("Wisdom University"), which seemed to me to be the most interesting place in Italy scientifically, apart from having an irresistible name. It was in Rome that the great Italian theoretical physicists were working, such as Gianni Jona-Lasinio, Giorgio Parisi, Nicola Cabibbo, Luciano Maiani and many others. From the director of the Department I obtained a table in a basement, where I spent a few years absorbed in the development of the new theory, ignored by all the colleagues. When the money of the scholarship was exhausted I did not succeed in obtaining any other financial support. Nicola Cabibbo, then the director of the INFN, had heard about my results in the USA and tried to arrange a contract with the INFN, but shortly after the political balance in the INFN changed and nothing came of it. I saved on everything to be able to live, and asked my father for help. My father, in spite of everything, believed in my scientific passion, helped me, and I am forever grateful to him for this.

It was a hard time of apprenticeship: I hoped to grow as a physicist, but my professional life seemed to have come to a halt without prospect. The possibility of getting a job at a university seemed remote, especially because I was working on a subject that in Italy almost nobody was interested in. There were moments of distress.

Then one day the telephone rang and the director of the department of physics of an American university asked me whether I was interested in a post as a professor. It was at the university of Pittsburgh, where Ted Newman worked, one of the greatest scientists in general relativity.

At first, the idea of living in a big and boring American city like Pittsburgh did not thrill me. My first answer on the phone was no thank you. It needed the intervention of Enzo Marinari, a dear friend quite wiser than me and a physicist in Rome, to open my eyes and to make me understand that I should not prefer to be unemployed in Italy rather than Professor in America. If I wanted the freedom to study what interested me, this was the opportunity.

At Pittsburgh I spent ten years, working with Ted and many other colleagues, devoting myself to many diverse problems, but above all developing loop theory.

Philosophers

One of the best surprises I found in Pittsburgh is the *Center for the History and Philosophy of Science*, perhaps the most important centre for the philosophy of science in the United States. It is an extraordinary institute where all kinds of visitors pass by and where one comes across all kinds of ideas. Ever curious and always fascinated by philosophy, I started to attend seminars and conferences of the Center.

There, I could approach eminent philosophers strongly interested in science, such as Adolf Grünbaum and John Earman. They were interested in spacetime problems and were eager to discuss these problems with a physicist. For me it was a great widening of my horizon and it linked up with the interests of my younger days. A lively dialogue was born, that taught me ideas and perspectives which turned out to be essential for my work as a physicist.

Entity or relation?

A typical issue on which a theoretical physicist has, today, something to learn from philosophy is the problem of the nature of space. In the western philosophical tradition, the idea of Newton that space is a very special *entity*, which exists also when nothing else exists, is not the dominant point of view on what space is. From Aristotle to Descartes, space has been more often described as a *relation* and not as an entity. This means that space does not exist if there are no objects. Space is the relation between dynamical objects, like a marriage is a relation between two individuals: there is no marriage without the two individuals. More precisely, space is the relation of *adjacency*, namely of being in touch, between objects.

Newton introduced space as an entity "within which" things exist, because this was necessary for his dynamical theory. He was able to show that this point of view about space is very effective for science. But the philosophical debate about the concept of space as an entity and the concept of space as a relation has continued through the centuries, furnishing scientists like Newton and Einstein with a theme for reflection and inspiration, and it has not exhausted its potential.

I am convinced that today it is necessary to reflect again on this sort of problems, in order to understand the quantum properties of gravity. I think that a complete theory of quantum gravity will only be constructed by abandoning the Newtonian idea of space as an entity, and interpreting general relativity in this way. Space is not an entity within which things are located: the space-entity does not exist. What exist is the gravitational field together with the other fields. In loop quantum gravity, the loops are quanta of the gravitational field and it is their relationships which constitute space.

More in general, I am convinced that today the dialogue between science and philosophy is vital. In the past, this dialogue has had a most important role for the development of science, especially at the moments of major conceptual evolution of the theory. Galileo and Newton, Faraday and Maxwell, Bohr, Heisenberg, Dirac and Einstein, to mention only the most important examples, were nourished by philosophy and would never have succeeded in accomplishing the gigantic conceptual steps they have, without this philosophical education. This is evident from their writings, in which truly conceptual and philosophical problems play an essential role: they suggest problems and yield new perspectives.

In the course of the second half of the twentieth century, fundamental physics has very much kept its distance from this dialogue with philosophy. The reason above all lies in the fact that the problems that were at hand during these decades had a technical rather than a conceptual character. Quantum mechanics and relativity had opened new worlds. The priority was then to study the consequences and the possible applications of these theories and the new worlds opened by them: atomic physics, nuclear physics, particle physics, condensed matter physics, astrophysics and cosmology. All these new sciences could be developed on the already established conceptual bases of quantum mechanics and general relativity. Today, on the other hand, in seeking to combine these basic theories, their conceptual foundations need to be addressed once more. Physics is once more concerned with foundational problems, and therefore a greater philosophical awareness is once more becoming essential.

This is also true from a methodological point of view: a scientist orients his own research on the basis of epistemological ideas. He might be more or less aware of them. Very often to be aware of your own assumptions is far better than to be guided by methodological prejudices of which you are unaware.

I also think, the other way around, that philosophy should not keep itself too distant from front line theoretical research in science. Scientific thinking is a "burning" way of thinking, in continuous evolution, and it raises fundamental problems.

Anglo-Saxon philosophy pays much more attention to active contemporary science than the philosophy of science of continental Europe. Through my Italian education I often feel closer to some continental European philosophy than to its Anglo-Saxon counterpart and since I came back to live in Europe, I have often tried to find again the dialogue I had in the United States with various philosophers of science. It is not impossible here, for instance I have opened an extremely fruitful dialog on quantum mechanics with philosophers such as Michel Bitbol of the École Polytechnique in Paris, and Federico Laudisa, in Milan. But most European philosophers look at science with suspicion.

The reciprocal suspicion between philosophers and scientists keeps the cultural divide between humanities and sciences wide open. I think this is short sighted: the culture of a civilization, including ours, is the ensemble of its knowledge and its ideas. It is stirring them *all* that knowledge progresses.

Grains of space and area spectrum

During the years I lived in the United States I returned to Italy every summer. Often Abhay Ashtekar and Lee Smolin accompanied me. They had become my friends and principal collaborators. We worked together in Italy and for them it was also a holiday. It was in Italy, during one of these periods, that we realized that to describe the macroscopic world, it was sufficient to superpose a great number of solutions, each made of a single ringlet. We understood how to put together a large number of individual loops and "weave" a classical macroscopic space with these loops, in the same manner in which a thread can be weaved into a smooth T-shirt.

Figure 3 illustrates a schematic model of this fine-structure of space: a tangle of loops. I constructed the model in those years to illustrate the idea, by making a round of all the ironware shops in Verona to buy all key rings that were available.

Figure 3: The first image of space suggested by loop theory. At the smallest scale space is a collection of small ringlets.

Another important step in the development of our theory was also reached in Italy, during a visit by Lee, towards the middle of the nineties. There was a technical aspect of the theory that we still could not understand: mathematically it should be possible for these loops constituting space, to *intersect*, to pass through one another at certain points. What did these intersection points mean?

We were dealing with a rather standard quantum mechanical calculation when we stumbled upon the clarification of the meaning of the intersection points. In quantum mechanics many quantities are "quantized". This means that they can only assume certain *discrete* values and not just any value. The energy of an ato m, for example, cannot have arbitrary values, but only certain special values that are called energy levels of the atom. To calculate the values that a quantity may have one uses a mathematical technique which is called “calculation of the spectrum of an operator”. Lee and I were interested in a particular physical quantity: the volume.

What is a volume? It is the measure of how much space there is. The volume of a room is the quantity of space there is in the room. Since space is the gravitational field, the volume measures the gravitational field. As we were dealing with a quantum theory there was the possibility that the volume had discrete values. The calculation turned out to be complicated. We got confused, but eventually we succeeded in solving it, thanks also to the help of a great English mathematician, Roger Penrose, whom we consulted when we were most confused, because certain mathematical objects he had introduced long ago, called spin networks, were appearing in our calculations.

The result of the calculation turned out to be that volume really is not a continuous variable; it is discrete, and hence that space is constituted of quanta of volume, or quanta of space. But this is not all: we also realized that these quanta of volume resided exactly on the *intersections* of the loops. In other words, the volume is composed of quanta, of finite grains of space, and the intersections of the loops represent precisely these grains of space. They are *the* grains of space we were looking for from the start.

These results changed our initial picture of quantum space somewhat. The intersection points became more important than the lines themselves: we no longer spoke of a collection of *loops* intersecting at points, but rather of a collection of *points* interconnected by lines, that is of a *network*. As these networks are mathematically characterized by certain half-integer numbers, or spins, associated to the lines, the networks were also called "spin networks". In fact, "spin network" was the name that Roger Penrose had assigned to this peculiar kind of networks, that he had earlier studied motivated solely by his intuition about quantum space.

The image resulting from quantum space is remarkable: the points of a spin network are the quanta of space. The lines connecting the points represent the spatial relationships. That is, they express which grain is near to which other grain. This is illustrated in Figure 4.

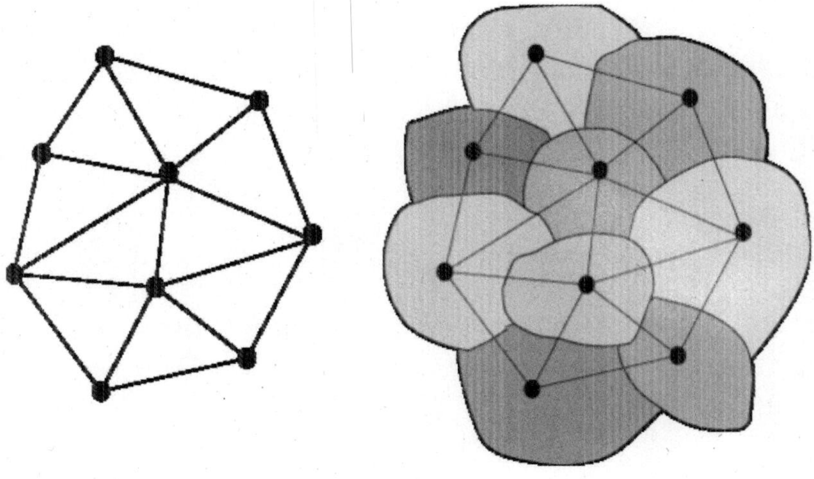

Figure 4: A"spin network" (left) is formed of quantum Faraday lines of the gravitational field; the intersection points, indicated by black dots, are the "nodes" of the network. They represent "grains of space" (right). The segments of the network represent the neighbourhood relations between the grains of space.

When we say that the volume of a room is, for example, 100 cubic metres, in reality we are *counting* how many grains of space, or rather how many "quanta of the gravitational field" there are in the room. Obviously these quanta are very small. In a room of 100 cubic metres their number is a figure of about one hundred digits. The calculation of the volume spectrum yields exactly the values of the volume that can be observed.

The same calculation can also be made for the area. Thus, one calculates also the "area spectrum". Loop theory predicts that if we measure an area accurately we would not get any arbitrary number but only one of the numbers of a list that can been obtained from a calculation. When we say the area of a page of this book is, let us say, 200 square centimetres, we are actually counting the number of lines of the network, that is of loops, that cross the page. Across the page of this book their number forms a figure of about seventy digits. This is illustrated in Figure 5.

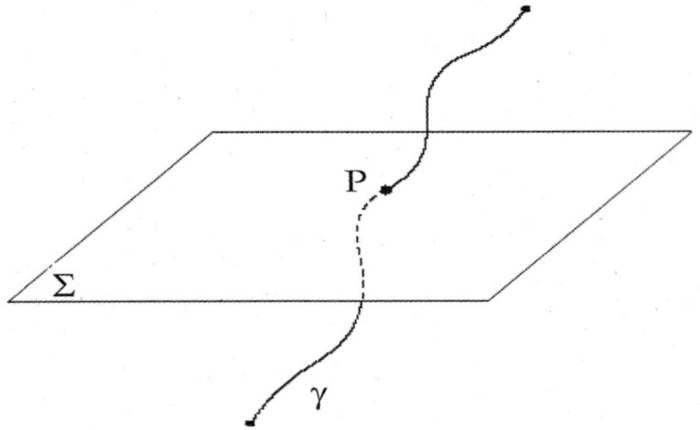

Figure 5: A surface Σ is crossed by a *loop* at the point P. The area of the surface is determined by the number of loops that cross it. One page of this book is crossed by about 10^{70} *loops*, that is, by a figure of *loops* consisting of about 70 digits.

The theory therefore predicts with precision a set of numbers representing the possible results of very precise area and volume measurements. Unfortunately, with today's technology one cannot yet verify these predictions. But it is very important that a theory gives precise predictions and, at least in principle, is verifiable. If it doesn't, it is not yet a real scientific theory. To this day, loop theory is the only theory of quantum gravity that furnishes an unambiguous and articulated set of predictions that are in principle verifiable.

The spin networks give a mathematically precise description of the quantum structure of space.

More precisely, since we are dealing with quantum mechanics – and hence with probability – the theory is formulated in terms of the probabilities associated with these spin-networks. One has to imagine that these spin networks constituting the world are fluctuating, vibrating and teeming like the black and white points of an old TV without antenna. The mathematics of the theory describes this swarming of the spin-networks.

This is a precise mathematical realization of the intuitive idea of John Wheeler according to which space on a small scale is no longer continuous.

I remember with great emotion the day when John Wheeler, the grand old man of quantum gravity, sent me a note full of affection and enthusiasm for our results and invited me to Princeton to expound our theory. In Princeton, he came to see me early in the Bed and Breakfast where I had found lodging. We had breakfast together and he then took me for a long walk on the campus. I told him the results of our calculations, while he told me his incredible stories: Bohr, Einstein, the atomic bomb… "The day Einstein arrived from Europe to the United

States escaping the Nazi, Carlo, I took him for this same walk on the campus..."

Why the vicinity of those who have left the strongest traces in our thinking causes such an emotion in us? They certainly are men and women like us, with the same weaknesses and the same limits. Still, the fascination of their ideas gives them an aura that enchants us. They have opened paths that we have the privilege of being able to follow, and this raises in us admiration, gratitude, and affection.

John talked with a soft voice. He was old and weak, but his interior energy was as intact. He defended his participations to the horrific adventure of the atomic bomb against my radical pacifist objections. When I showed him my image of the structure of space (see Figure 4 above), he smiled like a small boy and went fetching an extremely similar image that he drew many years earlier, and which is one of his books...

Theoretical applications

Today loop theory is studied by a great many research groups scattered around the world, that have developed it in many directions, some of which I do not follow anymore. The theory has applications in various fields, for example in the study of the properties of the black holes, in particular their thermal properties, and in cosmology, to study of the Big Bang – the very first moments of the life of the Universe.

The application of loop theory to black holes concerns a strange discovery made in the seventies by Stephen Hawking, the theoretical physicist who became famous because he has been able to continue his work as a scientist although he has been forced by a terrible disease to live in a wheel chair, and to communicate only by means of a computer he could operate with his fingers. Stephen discovered that black holes are "hot", that is, they behave exactly like hot bodies: they emit thermal radiation at a certain temperature. We know that in general objects are hot because their microscopic constituents move around. A piece of hot iron is a piece of iron in which the iron atoms vibrate rapidly around their equilibrium positions. If a black hole is hot, what are its elementary vibrating "atoms"? Loop theory offers a precise answer to this question. The elementary "atoms" of a black hole which vibrate and are responsible for its temperature, are precisely the individual loops sitting on the surface of the black hole. Using this theory one has been able to understand the result of Hawking in terms of the microscopic "vibrations" of the loops, and to calculate explicitly the temperature predicted by Hawking.

The application of loop gravity to cosmology is more complicated. Near the Big Bang the Universe was very small. In some sense it was only made of few grains of space. In such a situation, one cannot approximate its evolution by a continuous space, as one does when the universe is big. One must take granularity explicitly into account and therefore use the equations of loop theory. In this way one arrives at a description of the instants immediately after the Big Bang and even of the Big Bang itself. Remarkably, one finds that the equations of loop quantum gravity do not break down at this point, as do the Einstein equations. Therefore, the equations are capable of describing the Big Bang itself.

What do the equations show? What happens "before" the Big Bang? In my opinion it does not make sense to ask what has happened "before" the Big Bang: it is like asking what there is on the surface of the earth beyond the North pole. As I will discuss in the next section, "time" is a concept that only makes sense macroscopically. It looses its meaning in extreme situations such as near the Big Bang. We have to learn how to think the world without thinking at time. If we learn this, the problem of what happens "before the Big Bang" evaporates.

A number of other tentative applications of loop gravity regard astrophysics. For instance, the propagation of light over the granular space predicted by loop theory has been studied, and some indications that this propagation might be affected by this small scale structure in an observable way have been found. To understand how, recall that when light propagates in a crystal, the different wavelengths, namely the different colours, travel at slightly different speed. The microstructure of space might affect the propagation of light in the same manner. If so, we should receive the different components (with different colour) of the light of distant stars with small time delays with respect to one another. This may be tested and precise observations are under way. The problem is that this effect is difficult to compute within the theory, and therefore it is not completely clear what the theory really predicts.

Other applications are being developed right now, and I am not up to date with all of them. It is important to say, however, that so far none of these applications of the theory has produced a precise prediction that has been verified experimentally. This is true for all tentative quantum theories of gravity. Therefore loop theory, in spite of these beautiful results and in spite of the interest that the theory arouses, for the moment is only a *hypothetical* theory. It is a *possible* solution of the problem, but until we can perform measurements confirming its quantitative predictions, we will not know whether it is the *correct* solution. Keeping this distinction clear is, unfortunately, something which is too often forgotten by scientists writing popular works. Even more by scientists talking on TV shows.

A difficult aspect of a work such as mine is that on the one hand there is the enthusiasm to formulate a new theory: one may be on the right track to understanding a new aspect of the world; but on the other there is the frustration of the risk of working a whole life on theories that in the end might turn out to be wrong.

Time does not exist

Until now I have spoken only about space. The moment has come to speak about time.

Special relativity

Some ten years before he discovered the theory of general relativity, Einstein had already understood that space and time are not separate entities, but rather two aspects of the same entity. This discovery is called *special relativity*. More precisely the discovery of Einstein is as follows.

We are used to thinking that two events (the arrival of Columbus in America and the death of John Lennon, for example) are always ordered in time, that is, one happens before and the other happens later. We are used to thinking that time is something universal, and so it makes sense to ask what is happening *precisely now* in some part of the universe. Einstein understood that this is not the way nature works.

The most visual way to illustrate the relativity of time is the effect called "twin paradox". Two twins travel with great velocity far away from each other, and when they meet again they have different ages. One is older, one is younger. This is called a paradox, but it is not a paradox; it is simply the consequence of the way the world is made. The only paradoxical aspect of the situation is that we are not used to observing these phenomena. Thus they seem strange to us. But it is true: precise experiments have been performed (not with twins but with identical and very precise clocks put on board fast airplanes) and it has been verified repeatedly and with great precision that the world is made just as Einstein had understood: the two identical clocks exhibit different times when they meet again.

The point is that when two events happen in positions sufficiently far away from one another, it makes no sense to say which of the two happens first. It makes no sense to ask what is happening *precisely now*, on the Andromeda galaxy, for example. Time does not flow in the same way everywhere. We have our time and the Andromeda galaxy has its own time, and in general the two times are only loosely related.

The only thing one can do is to exchange signals. But signals will need six millions of years to go back and forth between Andromeda and here. Imagine an extraterrestrial who sends us a signal from Andromeda, a certain day, the first day of the year 1,000,000 in Andromeda Calendar. We receive this signal today and reply immediately. The extraterrestrial will receive our reply a certain day, say the last day of the year 7,000,000 in Andromeda calendar. We can say that the day at which the extraterrestrial has sent his signal (his 1,000,000) comes *before* today and the moment at which he receives the reply (his 7,000,000) comes *after* today. But between these two dates, there is are six million years. Which date on Andromeda is today? There is no answer. There does not exist on Andromeda a particular day that corresponds to "precisely today".

All this means that we must not think of time as if it there were a "cosmic clock" that marks the life of the universe. We must think of time as a something local: every object in the universe has its own time flow. The way in which everybody's time is put together when the objects meet or exchange signals can be precisely described, and was described by Einstein. To do this, one does not speak, in the description of the world, of "space" and "time", but of a union of these two called "space-time".

All this is known since more than a century (the work of Einstein in which he clarified all this was published in 1905); how come it is not more widely known? It should not come as a surprise that something we have known for a century has not yet become common knowledge to all people. The same thing has happened to many great conceptual revolutions, such as the Copernican revolution. Even a long time after the discovery of Copernicus, many people were still convinced that the sun was revolving around the earth and not the other way around. At the time of Galileo, almost a century after Copernicus book, many people, including rather powerful ones, were still quite sure of the undisputable truth that the earth is the centre of the universe... Research moves forward and does not wait, at each step, till all follow closely behind.

The meaning of time

I have explained above in which sense one can say that space does not exist anymore in quantum gravity: only the gravitational field exists, which is made of probabilities of elementary quanta, or grains, connected in networks. "Space" is just he network of these grains. Now, if time and space form a single entity, the non-existence of space implies also the non-existence of time.

Time does not exist: it is necessary to learn to think about the world in non-temporal terms. This is difficult, because we are accustomed to thinking about time as something in itself, which is flowing. But what does it mean that time does not exist?

In most equations of classical physics time occurs. It is the variable indicated by the letter "t". The equations tell us how things "change in time" and allow us to predict what will happen at some future time if we know what happened in the past. More precisely, we measure variables, for instance the position A of an object, the amplitude B of an oscillating pendulum, the temperature C of a body, etc. and the equations tell us how these variables A, B and C change in time. That is, equations determine the functions A(t), B(t), C(t) and so forth, that describe the change of these variables through time "t".

Galileo Galilei was the first to understand that the motion of objects on earth could be described by equations for the time-dependent functions A(t), B(t), C(t), and to write down the first equations that these functions had to satisfy. The first law of terrestrial physics found by Galileo, for example, described how an object falls when it moves down an inclined plane.

To discover and then confirm this law, Galileo needed two measurements. He had to measure the position of the object along the inclined plane, and the time t. He therefore needed an instrument to measure the time, that is, he needed a clock. In the period in which Galileo lived, there were no precise clocks, but Galileo himself at an early age had found a way to construct precise clocks. He had discovered that the oscillations of a pendulum all have the same duration. Thus one could measure the time by simply counting the oscillations of a pendulum. It seemed an obvious idea but it required a Galileo to find it. Before him nobody had thought of it; that is what science is like.

But things really are not so simple. Legend has it that Galileo had this intuition in the magnificent cathedral of Pisa while observing the slow oscillations of a gigantic candelabrum, which is still hanging in the cathedral today (the story is false, because the candelabrum was suspended there many years after Galileo's discovery, but it is a nice story). The scientist observed the oscillations of the chandelier during a religious service which evidently did not interest him very much, and he measured the beats of his pulse. With excitement he discovered that there was an equal number of beats during every oscillation. He concluded from that, that the oscillations all had an equally long duration.

The story sounds nice, but a slightly more careful reflection leaves one perplexed, and this perplexity forms the core of the problem of time. The perplexity is the following: how did Galileo know that *his pulse beats* all had the same duration?

A few years after Galileo doctors began measuring the pulse of their patients using a clock, which is essentially nothing else but a pendulum. Thus, one uses his pulse to be sure that a pendulum is regular and then uses the pendulum to be sure that the pulse is regular. Isn't this a vicious circle? What does it mean?

It means that we never measure time itself. We always measure physical variables A, B, C... (oscillations, beats, and many other things), and we always compare one variable with another one. Hence we measure the functions A(B), B(C), C(A)... and so forth. Yet it is *useful* to assume that there exists the variable "t", the "true time", that we can never measure directly, but which lies beneath everything. We write all equations for physical variables with respect to this *unobservable* "t". These equations that tell us how things change in "t" (how much time one oscillation requires and how much time each heartbeat takes). From this we can calculate how the variables change with respect to one another (how many beats in one oscillation) and we can compare these predictions with what we observe in the world. If the predictions are right we conclude that the whole complicated scheme is good, and in particular that it is useful to use the variable "t", even if we can never measure it directly. In other words, the existence of the variable time is an assumption, not the result of an observation.

Figure 6: Galileo Galilei discovers that the oscillations of a pendulum all are equally long, by counting the beats of his pulse during the slow oscillations of the chandelier of the Cathedral in Pisa.

It was Newton who clarified all this and set up this scheme. In the first section of his great book, the *Principia*, Newton asserts explicitly that we cannot measure the "true" time "t", but that if we assume it exists it gives us the possibility to construct an extremely efficient scheme to understand and describe nature.

So, I now return to our own days, to quantum gravity and to the significance of the assertion "time does not exist" that I stated earlier. The meaning of this assertion is simply that the Newtonian scheme no longer functions, when we are dealing with much smaller things. It was a good scheme, but only for macroscopic phenomena.

If we want to understand the world in a wider sense, if we want to understand it also in regimes that are less familiar to us, we must give up this scheme because it does not function any more. In particular the idea of a time "t" that flows on its own, and with respect to which the rest evolves, is no longer an effective idea. The world cannot be described by equations of evolution in the time "t".

When a physics student meets this idea for the first time, he usually panics. How can we describe the world without equations in which the time variable "t" does not appear? The answer is that in reality the time variable is not truly necessary to describe what we see. In fact, the time variable "t" is actually a "trick", codified by Newton, to describe something else. What we have to do to avoid using "t" is simply to restrict ourselves to list the variables A, B, C... that we *effectively observe*, and establish relations between these variables. We have to write equations for the functions A(B), B(C), C(A)... that we do observe; and not for the functions A(t), B(t), C(t), that we do not observe. In the example, we will not have the pulse and the pendulum that both evolve in time, but only equations that tell us how the one and the other may evolve with respect to one another. Not "how many beats per second and how many oscillations per second"; but only "how many beats per oscillation".

We are dealing here with a simple change of language for describing the world, but from a conceptual point of view the leap is gigantic. We have to learn to think about the world not as something that changes in time, but in another way. At the fundamental level there is no time. This impression of "time flow" that we all have is only an approximation that is only valuable for our macroscopic scales, and derives only from the fact that we observe the world only in a very coarse way. The elementary structure of physical reality is timeless.

A novel image of the world is taking shape: a world without space and without time. The space where the world "inhabits" and the time "along which" things evolve might soon disappear from our fundamental description of the physical world, in the same manner in which notions such as "the centre of the universe" have disappeared in the past.

Is it really possible to think the world in a fully a-temporal way? Can we really think the world without time, and get used to this? I think that the answer is yes, but it is a radical change in the structure of our thinking. Maybe as radical as the psychedelic dreams of my adolescence. Only the future will tell us if this is the right direction to go, in order to better understand this surprising and colourful world which is around us.

Return to Europe

An important part of my work has been to study the technical and conceptual problems raised by a theory without time. One of these problems is the following: if time does not exist at the fundamental level, what about the time we do perceive as the time that flows; what could that be?

Towards the end of the nineties I studied an idea for a possible solution to this problem, an idea which had an important effect on my life and brought me back to Europe.

Time is an effect of our ignorance

The world is complex: there are billions of particles and even more variables that describe the fields. We rarely have complete control over *all* the variables of a problem. When we do, we can verify that the system is governed by dynamical equations in which, at the fundamental level, as we have seen, time does not appear. However, most of the time we measure only a tiny fraction of the innumerable variables that characterize the system. For example, if we study a piece of metal at a certain temperature, we can measure its temperature, its length, its position, the speed at which the piece of iron is moving, but not the microscopic movements of each of its little atoms which, as we know, are responsible for the temperature. In these cases we not only use the equations of dynamics to describe the physics of the object, but also those of *statistical mechanics* and *thermo-*

dynamics. These theories allow us to make some predictions in spite of the fact that we don't know the exact movement of *all* the microscopic variables.

Now, let me come back to the macroscopic emergence of time: the idea is that time appears only in this statistical, or thermodynamic context. That is to say that *time is an effect of our ignorance* of the details of the world. If we had complete knowledge of all the details of the world, we would not have the sensation of the flow of time.

I have worked a lot on this idea and on the possibility of making this idea concrete by giving it the appropriate mathematical form. The point is to show how typical phenomena related to the flow of time can emerge from an a-temporal world, when statistics is involved. I was very happy with my results, but the reaction of the scientific community has been one of perplexity.

Then one day I was in Cambridge, England, at the Newton Institute, which is one of those wonderful institutions where scientists from all over the world are invited to spend some time, with the sole purpose of having the opportunity to meet colleagues and to exchange ideas. The somewhat pretentious atmosphere of Cambridge, where the news that Britain has lost its empire hasn't yet arrived, wasn't too congenial to me, and the time I was spending at the Institute began to look a bit wasted. But that evening I was sitting, by chance, at the dining table next to an extraordinary personality: Alain Connes.

As one of the greatest living mathematicians, Alain has received the most important international awards in his field. But when we started talking, he looked like an enchanted and passionate boy, with the enthusiasm of an adolescent mixed with an astonishing intelligence, a volcano of ideas, not only in the field of mathematics but also in physics, a field in which he has obtained surprising results. We engaged in conversations about this and that and drank several glasses of wine. At a certain point we touched on the issue of time, and Alain dropped the sentence: "I have an idea about how time emerges, but nobody has taken me seriously". I pricked up my ears and asked for details. I had to insist, because Alain was not very willing to go into technical details, but at the end he began to explain his idea, sketching diagrams with his fork on the tablecloth and throwing crumbs of bread into the air as an explanation.

After some initial confusion I realized that the idea he was illustrating was exactly the same as the one I had worked on. I ran into my office in the upper floor, and returned to the dinner table with my two published works on the subject. We had used very different mathematics, but Alain quickly realised that mine was a special case of his.

When a scientist formulates an idea, he usually tends to believe that his idea his correct. If none of the others agrees, the scientist often will continue to believe he is right and that it is the others who are wrong – but he will have some doubts. If he then finds out that someone else has independently come up with the same idea, the temptation to believe that "us two" are right and all the others "don't understand", becomes irresistible.

Alain and I published a paper illustrating this idea, putting together the aspects understood by him and those understood by me. And I found in Alain a new, formidable friend, with a unique intelligence and intellectual passion. (Alain at the reception of his recent CNRS Gold Medal: "The career of a mathematician begins when he realizes that a subject he is studying is not presented in the right way: it starts with an act of rebellion...").

Having lived ten years in the United States, I started to have enough of it and I really wanted to return to Europe, as I often said. The collaboration with Alain was providential. The scientific world is somewhat like the court of the Sun King: it is enough to be a moment near the King for all the doors to open afterwards. In France, Alain is a strange and anarchist sort of King, but a King nevertheless. A few months after the appearance of our work I received a phone call from the *Centre for Theoretical Physics* of Luminy in Marseille: would I be interested to go and work there?

This time I had far fewer hesitations than ten years before when I left for America.

America

Leaving the United States, the country without a proper name ("America" is the name of the full continent), had its price: more than anything else, I have come to miss the daily vicinity of my colleagues in Pittsburgh, in particular of Ted Newman, an extraordinary figure, great scientist (he is the one who found the most general description of black holes, for instance), but, even more, an intensely human person, fiercely honest, capable of seeing and understanding the others in depth, and to smile in any situation. When I was annoyed by somebody's behaviour, or angry, Ted came in my office, let himself fell into a chair (he has the body and the manners of a big bear), and his smile was so sweet and full of soft irony, that all my anger vanished in a smile. Ted represented a model for me that I wanted to follow, a secure affection, a point of reference. I have the greatest nostalgia for our physics session, in front of the blackboard, both of us passionate and screaming, strange contrast with the softly spoken protestant style of the rest of the department.

Furthermore, I found it difficult to loose the precious vicinity of the *Center for the History and Philosophy of Science*.

Finally, it was difficult to leave the straightforward American simplicity, the confidence in people, the trust, the will to do things, which makes America great, so far away from the snares and the obstacles that in Europe make everything slow and difficult as if everything was immersed in syrup. In the United States there is a lot to learn for a European, and one may easily succeed in doing things that are immensely difficult in Europe. For the promising youths, especially, there are golden bridges in America, where in Europe everybody tells them to patiently wait their turn. Probably, without the opportunities offered by the United States, I would not have succeeded in staying on the path of science.

For a European, though, to stay in the United States is also difficult. The values are different, human relations are different. Too many aspects of American culture are intolerable for a European: the urban violence, the racial tensions, the death penalty, the absence of medical assistance for all and of social security, the abandonment of the weakest and the poor to their fate, the arrogance of money and power. The very idea of "social justice" is understood in almost exactly opposite manners, the two sides of the Atlantic. In the US, social justice means that anybody *who has the capacities* can reach the top, wherever he or she comes from. In Europe, the other way around, social justice implies the defence of the weakest, and therefore in particular those *who do not have the capacities*. The characteristic European solution for a depressed area is to put money in the local school; the characteristic American solution is to give money to the *best* of the children in the school, and take them out of that school. The European asks: yes, and the others?

Then there is the aggressiveness of American foreign policy, which is intolerable for many Europeans. In today's America, Europe sees reflected, and continuing, the horrors of its own colonial crimes. The negative interpretation is the European envy and the nostalgia for its own lost power. The positive interpretation is that the catastrophe of the last World War has left a bit of wisdom into the European soul, and the awareness that collaboration is better than confrontation. But in any case, the cynical hypocrisy with which the ideology of liberty and democracy is used as a camouflage for a brutal self-serving foreign policy is far too transparent for the European eyes. And the American certainty of its own moral superiority appears ridiculous to many Europeans (especially if they do not think too much at their own certainty of their moral superiority). The majority of the Europeans that go live in the United States feel strongly, perhaps after a period of enthusiasm, the nostalgia for the somewhat more drowsy and gentle Europe.

Moreover, the times began to change and already there were signs of the atmosphere of fear, of gloom and a certain kind of fanaticism, that today covers like a black cloak some aspects of civil life in the United States. It was time to leave again.

On my first trip to Marseille I was dazzled by the light, the sun, the crystalline green of the sea, the atmosphere of Mediterranean charm, old and timeless, of that extraordinary mixture of people in this old French city, and I fell in love with it immediately. The Center for Theoretical Physics in Luminy, part of the University of the Mediterranean, where I work today, is just out of town, close to the sea, in a harsh and splendid natural setting.

It is an ideal place to study. I live close to the sea, I have restored a small one-hundred years old wooden fishing boat, with its old Latin sail, the sail of the van Gogh boats, and when I don't work I venture below the white and wild cliffs where the seagulls fly.

Loops and strings

Loop gravity has developed today into a widely studied theory, which is being developed by dozens of research groups around the world. But it is not the only competitor, for a theory of quantum gravity.

There exist also other hypothetical theories seeking to resolve the problem of quantum gravity. The alternative that has been most studied is "string theory". This is a theory that supposes that the elementary particles are not particles but are little cords, called, indeed, "strings".

In spite of an apparent resemblance between *strings* and *loops*, the difference is substantial: strings are little cords that move around *within space*, unlike loops which *are themselves space*.

The advantage of loop theory is that it incorporates in a fundamental way what general relativity has taught us about the world, namely the fact that Newton's table-space, on which particles move around, does not exist. The advantage of string theory, on the other hand, it that its ambition is bigger, since it tries to provide an unified description of all interactions: a single formula that should summarize all of nature.

There are various reasons for which I do not like string theory. One is that the theory predicts an enormous amount of stuff, for which there is no evidence: many extra dimensions, super-symmetric particles, new exotic kinds of particles. Experimental evidence for all this has been searched, but without any positive result. Second, I think that string theory fails to incorporate the true discovery of general relativity: namely the fact that spacetime is just a field. In string theory, one starts by assuming that spacetime is fixed and given, and puts strings in it. Nowadays there are efforts to formulate string theory also without this "backgound" spacetime. But when it does so, string theory begins dangerously to resemble, and to face the same issues, as loop theory. Finally, I have the impression that the string theory community has been following a research path that has taken it too far away from what we know about reality, without sufficiently asking whether or not that path is the correct one.

The difference between string theory and loop theory is not just in the different physical hypotheses they explore. It is also that the two theories are the product of two different scientific communities, that see the problem of quantum gravity somewhat differently. String theory has emerged from the community of the high-energy physicists that studied the elementary particles. This community that has obtained spectacular scientific successes during the twentieth century, and is therefore proud of its own peculiar methods, in particular quantum field theory. This community tends to see general relativity not as a major revolution is our understanding of space and time, but rather as a peculiar funny theory, that should be tamed into the same schemes that have worked for the other theories.

Loop theory, on the other hand, has grown more from the community of the relativists, the specialists of general relativity. This is a community, clearly, far more aware of the specific physical meaning of general relativity, and less impressed, on the other hand, by the techniques of quantum field theory. For a relativist, therefore, quantum gravity is more about incorporating quantum field theory into the conceptual revolution of general relativity, than in taming general relativity into conventional quantum field theory.

It has been suggested that the two theories could end up being complementary, and finally to merge. It is a nice suggestion, even if, for the moment, is not substantiated by many solid result.

There are several other directions that are being explored, for instance Alain Connes non-commutative geometry, that I have found particularly interesting and on which I have worked for while, and Roger Penrose's twistor theory. I very much recommend Penrose's difficult but splendid popular book "The Road to Reality", for an overview of various points of view: it a huge and insightful fresco on everything we know about the physical world.

Whether one of these theories is the right one, and which one that will be, or whether in the end one may be able to combine them, at present nobody knows. Future will tell.

Established theories versus hypothetical theories

How will we know which theory is the good one? Ultimately, experiments will tell. But none of the existing theories of quantum gravity is complete and none has been shown to be fully consistent. All of them have holes and have many sides that are not yet understood. Therefore, even before the experiments, some theory might turn out to be internally inconsistent, or inconsistent with what we already know about the world. But at the end, we will begin to have confidence in a specific theory only when it will have proved itself right in predicting something new, when we will have tested it in the laboratory, or with astronomical observations.

I have made an effort to make clear that loop quantum gravity is a theory that we do not yet know if it will turn out to be correct or not. Unfortunately, this is not always done in popular science writing. Even less on TV. Too often scientists sell hypothetical theories to the large public, as if they were established theories. I have seen this often done, for instance, with string theory. I think this is a great mistake, because it questions the credibility itself of science. We scientists live out of public money and it is our duty to be fully honest in reporting what we know and what we do not know. We are paid to dream, but we must not sell our dreams for established realities.

The distinction between established theories and hypothetical theories is not always completely clear cut, but this does not mean that we can play with it. What we know today about the physical world concerns Maxwell's equations, Newton's mechanics, quantum mechanics, the so called "standard model of elementary particles" and general relativity. These are fundamental theories that are established, and function very well in their domain. These theories can be trusted: if we build a machine based on these theories, we can have extremely good confidence that the theory will be correct.

Everything that goes beyond this set of theories, such as loop gravity, strings, super-symmetry, extra-dimensions, multi-universes, holographic principle, and many other ideas we hear from scientists on the media, all this, for the moment, is just speculation. It might turn out to be right, of course: the established theories of today were speculative theories in the past. But it might also turn out to be wrong: it has happened very often in the past that a large number of physicists had enthusiastically embraced a speculative theory that later turned out to be a wrong direction.

To speculate is a necessity: it is the only way for finding something new. But to be blinded by our own speculations is naivety. I am afraid that today many of my colleagues are dangerously close to the boundary of this naivety.

Today fundamental science is in a state of confusion. There are beautiful ideas on the table, but not yet confirmed. Maybe one of these is correct, and we have just to verify it experimentally. Or maybe something is missing, and we need a new idea. A new idea that will be found by some young scientists who perhaps, like Einstein, can't even find a job right now. Or by a young woman, or man, who hasn't entered research yet. Maybe you, my reader?

Why study space and time?

There are several kinds of science. Physics is made of equations. One part of physics is concerned with *applying* these equations to understand this or that phenomenon. Another part, however, is concerned with *seeking* these fundamental equations: this is called *fundamental* physics.

Fundamental science is always motivated almost uniquely by curiosity. By the "desire to know". Without fundamental science the whole of science would not exist. Without Galileo and

Newton, whose pursuit was fundamental science, we would not be able to perform calculations to erect bridges and big palaces. Maxwell's wanted to know the nature of light. Without him and Faraday there would not be radio and television. Schrödinger wanted to unravel the deep reason of the periodic table of the elements. Without him and Heisenberg there would be no computers and little of modern technology. At the time these scientists developed their theories, future applications were unthinkable. It was not the applications that motivated them but the natural desire to know. Our civilisation is a product of curiosity.

But the value of fundamental science, in my judgement, goes even beyond its possible applications - now or in the future - and the immense material advantages it has given the world.

Science is a way of thinking that consists in the capacity of changing ideas about reality; in not trusting received ideas; in continuously discussing one's own schemes. Basic science, even more than being an exploration of the world, is an exploration of thought itself. It is an exploration of the possible modes of thinking.

What fascinates me most in science is the possibility of continually re-discussing ourselves and our very way of thinking. We cannot go exit our own thinking, but we can modify it from within. Everything can be discussed, even the most elementary assumptions such as our ideas about space and time. We redesign the world continuously. It is like the dream of my youth: continuing the research of the unknown world. Beyond reality, there is more reality to discover.

The nineteenth century developed an image of science that today cannot be defended: the idea that science is a collection of "true facts" that we have been discovered once and for all. The twentieth century has taught us that things are more complicated than this. We have realized that scientific knowledge keeps changing. Even what is probably the greatest and most successful of the scientific theories ever, Newton's theory, is, in a very precise sense, badly wrong. Newton's theory retains all its validity in certain domains and within certain approximations, of course: we use it everyday to build a bridge or design an airplane. But we have realized that the picture of the world offered by Newton's theory is partial. It does not allows us to understand the world more at large. A theory is not expected to be "definitely true", even when it is well verified by experiments. Today, scientists are deeply aware of this provisional character scientific knowledge (not all of them, unfortunately some always think that their last paper will be the Final Theory of Everything, sometimes even before any experimental confirmation!).

But then, if science is so uncertain, is it credible? How can we trust it, if we know that even the best theories are likely to be found wrong sometimes in the future? I think that the answer to this question is important, because science, and more in general, rational thinking, is today under attack from many sides.

Science is not credible because it is absolutely true. But because on a large number of problems, it is the best answer we have found so far.

It is precisely the fact that scientific knowledge changes, which makes scientific knowledge so strong. Science is based on doubt, and this is the opposite of most of non-scientific thinking. It is doubt that makes our knowledge advance, because what blocks us from understanding, most of the time, are wrong ideas that we hold. To understand that the Earth is round, we have to be ready to give up a cherished image of the world in which we previously believed, where the Earth was flat. To understand that the Earth moves around the Sun, we have to be ready to give up a cherished image of the Earth at the centre of the universe, in which we previously believed. To understand that time does not exist, we have to be ready to give up a cherished image of the world evolving in time, in which we now believe. It is a continuous process, that follows the slow decrease of the immense ignorance in which we are immersed. To be capable to increase our knowledge, to learn something more, we have to be ready too accept that something we believe to be true is not actually true. This is why doubt, uncertainty, is the root itself of knowledge.

And this is also the reason for which the worst enemy of knowledge is certainty.

The worst enemies of knowledge are those who are certain that they already know the truth. Those who would never accept their certainties to be questioned. Those who think that all the basic things to be known are already know. These people, unfortunately, are many, in Universities, Governments and Churches. I have never understood how so many people can be so sure of their ideas. Don't they see that their neighbour is equally sure of something different? How can so many people have the arrogance, the silliness, of believing that they are the *they* are the depositary of Truth? That God spoke to them and not to others?

Science is the result of the opposite attitude. Questioning encrusted assumptions, be ready to see the world with new eyes.

Knowing that what we know might be incorrect. Be open to change.

This does not mean that we can never know anything. To the contrary, science is the process itself through which we can ameliorate our understanding of the world, but study, attempts, and, especially, by dialog. Rational thinking is the idea that if we respect points of view different from ours, new ideas proposed, and different opinions, then we can discuss these different opinions, analyze them, criticize them, make them clash, until clarity and agreement is reached. What is very remarkable is that it may take time, but science does lead to agreement. There are no discussion anymore on whether the Earth is the centre of the universe or not. It has been a big discussion, now it is solved. The key is critical discussion of ideas, facts, theories and experiments. At the end, one way of thinking, one theory emerges as the best one. It will not be the final theory of the world, but it will be the best description of the world we can give, at the present stage of knowledge. Be open to continuous change is not the weakness of science: it is its strength.

The trust in the possibility of discovering something new, of changing, of constructing new things, has allowed civilisation itself to take shape. It has allowed man to get rid of the fears of our ancestors, to lose some prejudices, to abandon the structures of the past, to create a somewhat fairer world, to want to understand before judging and deciding. Science is born when certainties, myths and old values are questioned and discussed, and one looks at the world with new eyes, like a child which is still open to everything.

I think this is what the world needs, especially now. It seems to me that today the world risks to become a prey of fear and arrogance: fear for those different from us and the arrogance of belonging to the right party. On the one hand, fear is exploited equally by desperate people in the poor world and by governments in the rich world; both seek to strengthen themselves and to find legitimacy in this fear. On the other hand, the arrogance of fundamentalism – Christian, Islamic, Hindu... – is growing, as well as the arrogance of the rich world, which is so certain of its own form of justice and, at the same time, so frightened, that it believes it has the right to fill the world with its armies.

The seventeenth century in Europe comes to my mind, with the devastation of the Thirty Years War when Catholics and Protestants massacred each other in the name of Truth. Religious wars, violence, misery and opulence, opulence threatened by poverty. Each side convinced of being on the side of justice, massacring one another in the name of God, of Peace or of "our Country". Calling one another heretics, or terrorists.

In all this, science is obviously largely irrelevant. But perhaps, while being in the margins, it can still play a bit the role it played in the tragic European seventeenth century. The role of safeguarding the idea of a way of thinking that knows that its own point of view is not the absolute truth, a way of thinking based on serene rationality, willingness to listen to adversaries instead of trying to destruct them, and based on tolerance.

In the seventeenth century, intellectuals continuously crossed the borders of a devastated Europe. They nourished the dream that the rational comprehension of the world which the Greek world had introduced, but which had been lost for centuries, crushed by dogmatic truths, would come to life again. Through the books and the thinking of that group of scientists the modern world was shaped.

Also today men and women of science travel ceaselessly all over the world, although the difficulties to obtain visas and to cross borders are growing. They all feel they are citizens of the same small and inconsiderate planet.

Sante Di Renzo, who came up with the excellent idea of this series of interview-booklets, asked me what advice I would give to a young student who faces life. I have no real advice to give, as I have ignored most (but not all) of the advice I myself received, but if this little book will be read by a young student, my wish would be that he would continue to believe in his dreams, whatever they are, and would not allow himself to be intimidated by whoever preaches fear. Certainly one must know how to accept compromise, but only by believing one's own dreams does one have the possibility of doing something good for oneself and for the world.

What I have learned in science is that there isn't a single solid real world, understood once and for all. The world is always different from what we think: it changes under our eyes. We can dream of new worlds, seek them, set them up. The vision of the world we have inherited is the result of the rebellion of the previous, it is the result of their effort to think the new. Our realities are the realization of their dreams. We can continue our own rebellion, and dream the world afresh. If it works, it might increase or knowledge, or maybe make the world a bit better. If it doesn't, it is nevertheless a wonderful way for spending these brief years we have.

The scientific work I have done until now might lead nowhere: in the end it could turn out to be all wrong. Or it could be, as I dearly hope, a small step forward in the understanding of our colourful world. Either way, the endless hours spent in front of blackboards covering formulas, or in discussions with innumerable friends, colleagues and students of every country in the world, have been marvellous, and I feel very happy to have lived through them.

Today I am surrounded by youngsters fascinated, like I was more than twenty years ago, by fundamental research. They come to see me from all over the world, as I had done then. I talk to them, tell them what I know, hoping that among them will be one who, being better than me, reaches there where we were not capable to arrive. When they ask my opinion I strongly advise against undertaking a scientific research career in this field, like my professors had done to me. I tell them about the fierce competition for posts, about the difficulty of the subject, of the enormous risks of this difficult profession. But inside me I hope that they have the passion and the force to ignore all these cautions, to pursue their passion, and to believe their dreams.

* * *

I would like to take this opportunity to thank all the numerous friends with whom I have had the luck to work, and with whom the ideas described here have been developed. Among these, in particular:

Abhay Ashtekar, John Baez, Julian Barbour, John Barrett, Michel Bitbol, Mauro Carfora, Alain Connes, Louis Crane, Roberto De Pietri, Bryce DeWitt, John Earman, Jurgen Ehlers, Laurent Freidel, Hugo Morales-Tecotl, Giorgio Immirzi, Chris Isham, Ted Jacobson, Gianni Landi, Federico Laudisa, Jerzy Lewandowski, Renate Loll, Leonardo Modesto, Ted Newman, Daniele Oriti, Massimo Pauri,

Roger Penrose, Alejandro Perez, Jorge Pullin, Micheal Reisenberger, Philippe Roche, Simone Speziale, Lee Smolin, Massimo Testa, Thomas Thiemann, Bas van Fraassen, and John Wheeler.

Some recommended reading

Lee Smolin, *Three Roads to Quantum Gravity*, Oxford University Press, 2000. Excellent popular introduction to quantum gravity.

Carlo Rovelli, *Quantum gravity*, Cambridge University Press, 2004. Technical introduction to quantum gravity.

Greg Egan, *Schild Ladder*, Gollangz, 2001. Beautiful science fiction novel, inspired by loop theory.

Roger Penrose, *The Road to Reality: A Complete Guide to the Laws of the Universe*, Knopf, 2005. A great fresco of all we know about the physical world.

Sommario

Also available:

Carlo Rovelli
What is Time? What is Space?
*
Jacob D. Bekenstein
Of gravity, black holes and information
*
Paul Davies
One Universe or Infinite Universes?
*
Giuseppe Arcidiacono
Projective Relativity, Cosmology and Gravitation
*
Ignazio Licata (ed.)
Majorana Legacy in Contemporary Physics
*
Adalgisa Battistelli (Ed.)
Innovation in the trasformation of jobs and organizations

I dialoghI

SCIENZA

Sidney Altman	La rivoluzione dell'RNA
Giuseppe Arcidiacono	L'uomo, la vita, il cosmo
Tito F. Arecchi	Coerenza, complessità, creatività
Michael F. Atiyah	Siamo tutti matematici
Jesse H. Ausubel	La liberazione dell'ambiente
Francisco J. Ayala	Le ragioni dell'evoluzione
Jacob D. Bekenstein	Buchi neri, comunicazione, energia
Edoardo Boncinelli	A caccia di geni
Luciano Caglioti	La scienza tradita
Luigi Campanella	La chimica e oltre
Federico Capasso	Avventure di un designer quantico
Luigi Luca Cavalli Sforza	Il caso e la necessità
Alan Cromer	Visioni della scienza
Paul Davies	Un solo universo o infiniti universi? *(anche in ebook)*
Bryce DeWitt	Sopra un raggio di luce
Carl Djerassi	Dalla pillola alla penna
Freeman Dyson	L'importanza di essere imprevedibile
Irenäus Eibl-Eibesfeldt	Dall'animale all'uomo
Antonio García-Bellido	Vocazione e ricerca
Sir James Gates, Jr.	L'arte della fisica
Michael S. Gazzaniga	L'interprete *(anche in ebook)*
Jane Goodall	Cambiare il mondo in una notte *(anche in ebook)*
John Gribbin	Dalla scimmia all'universo
David J. Gross	L'universo affascinante
Umberto Guidoni	Il giro del mondo in 80 minuti
Margherita Hack	Una vita tra le stelle
Hermann Haken	Nel senso della sinergetica
Roald Hoffmann	Come pensa un chimico?
Massimo Inguscio	Fisica atomica allo zero assoluto
Giorgio Israel	Scienza e storia: la convivenza difficile
Arie Issar	Rifiorirà il deserto
Harold Kroto	Molecole su misura
Anthony J. Leggett	Il mistero della superfluidità
Vladilen S. Letokhov	Dalla Siberia alla scienza del laser
Jean-Marc Lévy-Leblond	Scienza e cultura

Alfonso Maria Liquori	Etica ed estetica della scienza
Luigi Lugiato	Divertirsi con la ricerca
Benoit Mandelbrot	Nel mondo dei frattali
Desmond Morris	Linguaggio muto *(anche in ebook)*
Igor D. Novikov	Il ritmo del tempo
George A. Olah	La ricerca di una vita
Francesco Paresce	Tra razzi e telescopi
Giorgio Parisi	La chiave, la luce e l'ubriaco
David Peat	I sentieri del caso
Arno Penzias	L'origine dell'Universo
Martin L. Perl	Tante domande, qualche risposta
Max Perutz	Le molecole dei viventi
Luciano Pietronero	Complessità e altre storie
Ilya Prigogine	Il futuro è già determinato?
Martin Rees	La lucciola e il riflettore
Carlo Rovelli	Che cos'è il tempo? Che cos'è lo spazio? *(anche in ebook)*
Dennis William Sciama	Questo bizzarro universo
Warren Siegel	Particelle, stringhe e altro
Stephen Smale	Matematica sulla spiaggia
Orazio Svelto	Il fascino sottile del laser
Giuliano Toraldo di Francia	In fin dei conti
Imre Toth	Matematica ed emozioni
Lucia Votano	La via della seta. La fisica da Enrico Fermi alla Cina *(anche in ebook)*
Frank Wilczek	La musica del vuoto
Richard N. Zare	Molecole e vita
Semir Zeki	Con gli occhi del cervello *(anche in ebook)*

Uomo e società

Joseph Agassi	La filosofia e l'individuo
Sadik Al-Azm	L'Illuminismo islamico
Emmanuel Anati	I segni della storia
Dario Antiseri	Idee fuori dal coro
Robert J. Aumann	I giochi dell'economia e l'economia dei giochi
Ermanno Bencivenga	Per gioco e per passione
Aldo Carotenuto	La mia vita per l'inconscio
Noam Chomsky	Linguaggio e politica *(anche in ebook)*
Daniel C. Dennett	Dove nascono le idee
Alfonso Maria Di Nola	Attraverso la storia delle religioni
Gianfranco Dioguardi	L'avventura della ricerca
Sante Di Renzo	La centesima pagina *(anche in ebook)*
John Dunn	Pensare la politica
Paul Ekman	La seduzione delle bugie *(anche in ebook)*

Christopher Freeman	La creatività scientifica nello sviluppo economico
Giorgio Galli	Il ritorno del rimosso in politica
Clive W.J. Granger	Economia di pace, economia di guerra
Adolf Grünbaum	La mia odissea dalla filosofia alla psicoanalisi
Allan J. Hobson	Sognare
Boris Kagarlitsky	Dove va la Russia
Ervin Laszlo	L'uomo e l'universo
Bernard Lewis	Uno sguardo dal Medio Oriente
Emmanuel Milingo	Una risposta al male
George Mosse	La nazione, le masse e la nuova politica
Ernst Nolte	L'eredità del nazionalsocialismo
Jaroslav Pelikan	Continuità e cambiamento della fede
Gianni Perrelli	Professione reporter
Franco Prattico	La lampada di Aladino
Jonas Rappeport	Nella mente di un criminale
Sergio Ricossa	Dov'è la scienza nell'economia?
Jonathan Riley-Smith	Al seguito delle Crociate
Pascal Salin	Liberismo, libertà, democrazia
Dominick Salvatore	Verso un'economia globale
Hideo Suzuki	Climatologia e storia delle religioni
Arthur Vidich	Cultura, economia e classi sociali nella politica degli Stati Uniti

ARTE E SPETTACOLO

Pier Augusto Breccia	La pittura come visione del profondo
Richard Brilliant	Un americano a Roma
Leo Gullotta	Mille fili d'erba
Federico Zeri	Caro Professore

LETTERATURA

Yehuda Amichai	Ogni uomo nasce poeta
Clara Sereni	Da un grigio all'altro

MEDICINA

Ciro Campanella	Dal profondo del cuore. Diario ed esilio di un cardiochirurgo *(anche in ebook)*
Gian Carlo Di Renzo	Il mio paziente non è ancora nato
Mirko Drazen Grmek	Malattie in agguato

DOCUMENTI

Giorgio Di Bernardo Nicolai	Nella nebbia, in attesa del sole
Milvia Spadi	Le parole di un uomo. Incontro con Primo Levi

IMPROBABILI

Guido Del Giudice	Io dirò la verità. Intervista a Giordano Bruno
Marco Margnelli	Gli stati modificati della coscienza

IN INGLESE

Jacob D. Bekenstein	Of Gravity, Black Holes and Information *(anche in ebook)*
Richard Brilliant	Roman Art, an American's View
Paul Davies	One Universe or Infinite Universes?
Carlo Rovelli	What is Time? What is Space? *(anche in ebook)*

In Bookstores

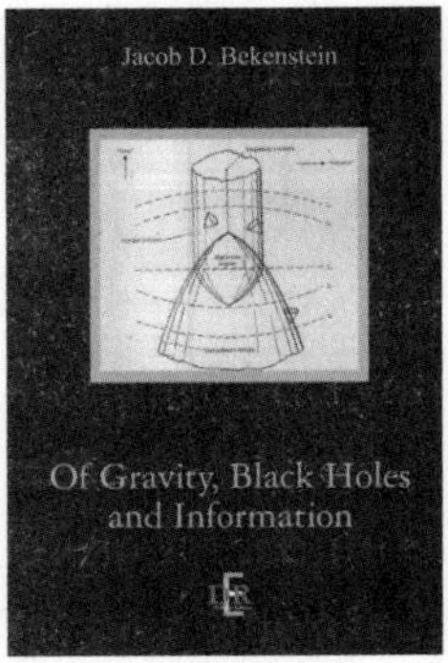

Black holes are among the strangest denizens of the physicist's world. Jacob Bekenstein tells here the story of how the thermodynamics of a black hole was discovered by him and Stephen Hawking, following John A. Wheeler's question about what would happen if a cup of hot tea were dumped into a black hole. He further describes ramifications of these thermodynamics, such as the mysterious information paradox and the holographic bound, which have deeply influenced theoretical physics in the last twenty years. These subjects are just some of the many adventures in physics which the author has witnessed and participated in. He describes the genesis and development of some of them in this book.

Jacob David Bekenstein was born in Mexico City. He received his Ph.D. from Princeton University in 1972 and then was a postdoctoral fellow at the University of Texas at Austin, and a faculty member and Arnow Professor of Astrophysics at the Ben-Gurion University, Israel before moving in 1990 to the Hebrew University of Jerusalem where he is the Polak Professor of Theoretical Physics. Bekenstein is a member of the Israel Academy of Sciences and of the World Jewish Academy of Sciences. He received the first prize of the Gravity Foundation of the USA in 1981, the Rothschild Prize in 1988 and the Israel Prize in 2005.

Some of his 100 publications: *Black Holes and Entropy* (1972); *The Energy Cost of Information Transfer* (1981); *The Fine Structure Constant: Is it Really Constant?* (1982); *Spectroscopy of the Quantum Black Hole* (1995), and *Relativistic Gravitation Theory for the Modified Newtonian Dynamics Paradigm* (2004).

In Bookstores

In this book, Giuseppe Arcidiacono develops the new Special Projective Relativity theory, based on the Fantappié group and the De Sitter Universe with constant positive curvature, which extends the Einstein Special Relativity theory on a cosmic scale, obtaining a group version of the Big Bang cosmology.

Giuseppe Arcidiacono (1927-1998) was Professor of Mathematical Physics at the University of Rome and at the University of Perugia (Italy). He was a researcher with Luigi Fantappié at the "Istituto Nazionale di Alta Matematica" of the University of Rome and head of the "International Center of Comparison and Synthesis".
His primary research interests were in the field of the mathematical theory of relativity, cosmology and the unified theory of matter and electricity.

In Bookstores

The author has a distinct talent for rendering his complex subject both interesting and accessible to his readers. It is this particular quality that has made him well known throughout the world.
The author shares with the reader his personal and intellectual flow of ideas from the beginning of his career to the present day.
We are convinced that the reader will find this volume both captivating and informative as the author guides us through some of the greatest mysteries still left to man.
Paul Davies is a scientist and writer whose career has taken him from investigating the deepest mysteries of the Universe to the problem of life itself. This wide-ranging but accessible book provides the best possible introduction to the man and his work.

John Gribbin

Paul Davies is Professor of Natural Philosophy at the University of Adelaide Australia. He obtained a doctorate from University College London in 1970 and he has also held academic posts at the universities of London, Cambridge and Newcastle-upon-Tyne.
He emigrated to Australia in 1990. His research interests are in the field of black holes, cosmology and quantum gravity including su-

perstrings, higher-dimensional black holes and quantum cosmology. Among his better known works are *God and the New Physics*, *Superforce*, *The Cosmic Blueprint*, *The Mind of God* and *The Last Three Minutes*. His latest books include a major work of popular science entitled *About Time*, and *Are We Alone?*, dealing with the search for extraterrestrial life.

In Bookstores

This volume focuses on the analysis of the beholder's active share in the reception of works of visual art, whereby their meaning is shaped and appropriated by viewers of the time of creation, and subsequently. Therefore, the study of artworks in and through history, their careful visual apprehension, and their exposition, both orally in teaching and in writing to a larger public, constitute a seamless web of interactive functions, fundamental to the role of art historian. In that capacity, the visual delight and stimulation offered to the eye and mind transcend personal experience, without which there cannot be any valid interpretative activity, and imposes on that experience the need to share it in a disciplined, well-founded manner with others.
Perhaps it is that motive that so energized Brilliant's writing: to be so attentive to the stories told to make sense of the world, and to the images of ourselves, provided by us in person and portrayed by artists that establish the realities of our social existence and make them worthy of consideration.

Richard Brilliant is Anna S. Garbedian professor in the Humanities and professor emeritus of Art History and Archaeology at Columbia

University. He has taught a variety of courses on Greek and Roman art, visual narrative, portraiture, and the theory of art history.
In the past forty years he has published thirteen books and more than one hundred articles and long reviews.
Trained as a classicist, lawyer, and art historian, in Yale and Harvard, he long ago moved beyond his base in Greek and Roman art to examine the effect of the historiographic tradition on the interpretation of artwork, to define narrative strategies in ancient and medieval visual imagery, and to investigate the interpersonal relationships in post-Renaissance portraits among the subject, the artist, and the viewer..

Finito di stampare nel mese di febbraio 2019
presso *Tipografia e Servizi Politano srls*
Via Casilina Vecchia, 147 - 00182 Roma